GENE BEHAVIOR IN BACTERIA EXPOSED

How the New Tool Revealing Gene Behavior in Bacteria was discovered and everything to know about it

JOHNNY MAX CARSON

This BOOK belongs to

____

____

From:

____

Signature/Date:

____

TABLE OF CONTENTS

INTRODUCTION

In the continuously changing environment of technological progression, creativity, and funding are inextricably linked, moving us toward a future marked by progress and transition. From healthcare to space exploration, the combination of innovative ideas and financial support is transforming sectors and our sense of what is possible.

This thorough investigation dives into the critical role of technology advancements and financial efforts, investigating how these dynamics affect diverse sectors. As we manage the complexities of technology advancements and financial support, a story of limitless possibility develops, suggesting a future in which ideas, driven by strong financing mechanisms, dictate the course of our collective journey.

Let us unravel the fabric of innovation and financing in healthcare, education, sustainability, and beyond, emphasizing their significant effect on how we live, work, and imagine the future.

CHAPTER 1

UNRAVELING THE COMPLEX LINK BETWEEN DNA REPLICATION AND GENE REGULATION

Understanding the processes of gene regulation is critical for decoding the complicated dance of life in the intricate realm of molecular biology. Pioneering research from the NYU Grossman School of Medicine and the University of Illinois Urbana-Champaign has revealed a fresh way to shed light on the complex interplay between DNA replication and gene regulation in bacteria. This discovery not only advances our understanding of microbial development but also shows promise in combating the alarming rise of antibiotic resistance.

DNA Replication and Gene Regulation

Bacterial infections, which take millions of lives each year, are an ever-increasing menace, compounded by bacteria' resistance to medications.

This problem revolves around bacteria's capacity to turn genes on and off in response to environmental changes, including drug exposure. Gene expression, or the translation of DNA into mRNA, regulates the creation of proteins required for the microbe's structure. Understanding how mRNA synthesis is controlled for each bacterial gene is critical for overcoming resistance, but previous ways of researching this regulation have been arduous.

Enter the new study, in which researchers discovered a fascinating link between DNA replication and gene control. Bacteria, like all living things, develop and proliferate by cell division. Before dividing, cells diligently duplicate their DNA to guarantee that each daughter cell obtains the entire set. This copying process is carried out by DNA polymerase, a molecular machine that reads and repeats each gene individually.

Revelation of Disruption

The study, published in the journal Nature, demonstrates that when DNA polymerase hits a specific gene during replication, the current transcription process is disrupted. This disruption, analogous to a ripple in a molecular pond, indicates the regulatory state of the gene. The continual duplication of genes during the bacterial cell cycle, as cells multiply and expand, provides an important tool for studying gene control. This finding provides a more efficient way than previous methods.

The transcription-replication interaction profile (TRIP)

The researchers used the term Transcription-Replication Interaction Profile (TRIP) to characterize the graphical depiction of these disturbances. TRIP, analogous to an electrocardiogram in medicine, reveals a succession of waves representing changes in mRNA quantity as genes replicate. TRIP patterns can be associated with unique regulatory properties. Genes under repression, where a protein prevents mRNA synthesis, show typical spiking TRIP patterns.

Implications and future directions

Andrew Pountain, the study's principal scientist, sees TRIP as a tool for identifying gene regulation across thousands of bacterial genes. Itai Yanai, the study's principal scientist, views TRIP as a tool for investigating how groupings of genes respond to environmental disturbances or changes, potentially providing important insights into bacterial behavior.

The researchers intend to expand their findings to particular TRIPs of genes involved in bacterial disease-causing processes. They believe that by decoding these patterns, they may be able to discover techniques to stop or slow illness development. As technology advances, enabling unparalleled capacities for tracking gene activity at the single-cell level, the researchers want to delve further into the complexities of gene behavior across distinct bacterial species.

In the continuous fight against antibiotic resistance, the relationship between DNA replication and gene regulation emerges as a critical battleground. The study's novel technique, which exploits the disruptive effect of DNA replication on gene transcription, gives insight into the regulatory complexities that regulate bacterial life. As the scientific community accepts the Transcription-Replication Interaction Profile, the opportunity to accelerate drug development and improve our understanding of microbial behavior becomes more real. This study not only represents a big step forward in deciphering the mysteries of bacterial gene regulation, but it also serves as a beacon of hope in the fight against antibiotic-resistant microorganisms.

CHAPTER 2

DECIPHERING THE GENETIC SYMPHONY

In the complicated orchestra of molecular biology, the synchronization of DNA replication and gene transcription is analogous to a symphony, directing cell life activities. A groundbreaking study done by experts from NYU Grossman School of Medicine and the University of Illinois Urbana-Champaign presents a unique concept: the Transcription-Replication Interaction Profile (TRIP). This profile, similar to an electrocardiogram for bacterial genes, promises to reveal the complexities of gene regulation, providing hitherto unattainable insights into microbial growth, responses to environmental changes, and possible antibiotic resistance tactics.

Understanding the TRIP

The TRIP notion arose from the researchers' investigation of how DNA replication affects gene transcription during bacterial growth. The molecular machine DNA polymerase orchestrates the critical phase of DNA duplication when cells divide. This study indicates that when DNA polymerase finds a certain gene during replication, it inhibits the transcription process. The resulting disruptions have a distinct signature—TRIP—and provide a graphical picture of the dynamic interaction between gene replication and transcription.

Analogous to an electrocardiogram

The researchers' comparison between TRIP and an electrocardiogram (ECG) in medicine is enlightening. TRIP produces a sequence of waves that indicate changes in mRNA quantity as genes replicate, similar to how an ECG shows the electrical activity of the heart. These waves, when plotted on a graph, form a visual depiction of the Transcription-Replication Interaction Profile, providing a full perspective into the regulatory dynamics of bacterial genes.

Linking TRIP Patterns and Gene Regulation

TRIP patterns function as a unique fingerprint for gene regulation. The study finds various TRIP patterns that correspond to distinct regulatory characteristics. Genes under repression, where a protein prevents mRNA synthesis, show typical spiking TRIP patterns. Deciphering these patterns provides researchers with significant information into the regulatory state of individual genes, making it a strong tool for understanding how bacterial genes turn on and off during growth.

Application in Disease Study

The researchers want to use TRIP in illness investigations, with a particular emphasis on genes involved in bacterial disorders. Investigating these genes' unique TRIPs might reveal important information about disease processes and new therapeutic strategies. Understanding the regulatory complexities indicated in TRIP patterns allows for the interruption or stalling of disease progression, opening up new possibilities for medicinal development.

Technological Advancements

The fulfillment of TRIP's promise was made feasible by technical advances in single-cell sequencing (scRNA-seq) and single-molecule fluorescence in situ hybridization (smFISH). These technologies allow researchers to observe gene activity in real-time at the single-cell level, revealing hitherto unseen insights into the dynamic nature of transcription and replication interactions.

Future Directions

As the study pioneers the use of TRIP, the research team plans to explore more individual TRIPs of genes from different bacterial species. The ongoing advancement of technology is predicted to allow for more precise examinations of gene activities, resulting in a better knowledge of microbial responses to environmental stimuli.

The Transcription-Replication Interaction Profile is a game-changing tool that will pave the way for future research into gene regulation in bacteria. TRIP is at the cutting edge of molecular biology innovation, with the potential to speed drug development, battle antibiotic resistance, and solve the secrets of microbial growth. As scientists conduct more research and technology improvements, the TRIP idea promises to not only unravel the genetic symphony of bacteria but also to inspire fresh therapeutic ways for combating infectious illnesses.

CHAPTER 3

TRANSFORMING APPLICATIONS ACROSS INDUSTRIES

In today's society, technology has emerged as a driving force, transforming how we live, work, and connect. Its disruptive impact stretches across several industries, altering processes, increasing efficiency, and opening up new opportunities. This thorough examination digs into the numerous uses of technology across sectors, demonstrating its varied effects on our everyday lives and the larger global scene.

1. Healthcare

Technology has brought forth a new age in healthcare, transforming patient care, diagnosis, and treatment methods. Electronic Health Records (EHRs) simplify information administration and enable seamless communication among healthcare practitioners. Telemedicine systems allow for remote consultations, which bridge geographical distances and provide access to medical experts. Advanced imaging technologies, artificial intelligence (AI), and data analytics improve diagnosis accuracy, opening the path for more tailored and precise medicine.

2. Education

Technology has broken down conventional educational barriers, providing novel tools for learning and cooperation. Virtual classrooms, online courses, and e-learning platforms democratize education by making it available worldwide. Immersive learning experiences are created using technology such as interactive simulations, augmented reality (AR), and virtual reality (VR). Learning management systems (LMS) provide efficient course administration and evaluation, resulting in a dynamic educational environment.

3. Business & Industry

Technology is a critical component of modern corporate operations, driving efficiency and agility. Cloud computing improves storage and processing capacities while promoting remote collaboration and scalability. ERP solutions streamline corporate activities ranging from supply chain management to financial transactions. Big data analytics improves decision-making by extracting important insights from large datasets, whilst cybersecurity solutions protect digital assets from developing threats.

4. Communication

The introduction of technology has caused a paradigm change in communication. Instant messaging, video conferencing, and social media platforms provide real-time worldwide communication. Voice over Internet Protocol (VoIP) technologies allow for cost-effective and high-quality voice communication. The development of smartphones and high-speed internet has changed the way we communicate information, resulting in a hyperconnected society.

5. Transportation

In the transportation industry, technology has given rise to smart mobility solutions. GPS navigation systems optimize route planning, resulting in shorter travel times and lower fuel use. Autonomous cars driven by AI algorithms provide safer and more efficient mobility. Ride-sharing services use technology to easily link commuters with drivers, resulting in a more sustainable and integrated urban infrastructure.

6. Entertainment & Media

The entertainment sector has gone through a digital revolution, with technology influencing content generation, delivery, and consumption. Streaming services provide on-demand access to a wide range of material, disrupting traditional broadcast methods. Virtual reality (VR) and augmented reality (AR) have transformed immersive entertainment experiences, blurring the distinction between the digital and physical worlds.

7. Environmental sustainability

Technology plays a critical role in tackling environmental issues. Renewable energy technologies like solar and wind power help to change the energy landscape toward sustainability. IoT (Internet of Things) devices track and optimize resource use, decreasing waste and encouraging conservation initiatives. Advanced analytics improve environmental studies by giving significant insights into climate trends and ecosystem dynamics.

Technology's applications are diverse, impacting every aspect of our linked environment. Its transformational potential is visible in healthcare, education, business, communication, transportation, entertainment, and environmental sustainability. As technology advances, its effect will only grow, promoting innovation, efficiency, and a dynamic global environment. Embracing technology's potential is not just a choice, but a requirement for navigating the intricacies of the current day.

CHAPTER 4

EXPLORING FUTURE DIRECTIONS

The future is a blank canvas that awaits the brushstrokes of invention and growth. The decisions we make now affect the course of our collective journey in every domain, from technology to healthcare, education, and sustainability. This inquiry dives into the fascinating opportunities and probable future paths across a variety of sectors, providing a glimpse into the ever-changing environment that lies ahead.

1. Technology and AI

The ongoing integration of artificial intelligence will shape the future of technology. AI, powered by machine learning algorithms, is poised to transform businesses by improving automation, decision-making, and problem-solving capabilities. Intelligent systems that can learn and adapt will play an important role in a variety of industries, including healthcare diagnostics and business predictive analytics. The rise of explainable AI strives to solve transparency and ethical problems, ensuring that AI systems are understandable and trustworthy.

2. Healthcare, Precision Medicine

The future of healthcare predicts a trend toward precision medicine, with treatments tailored to an individual's genetic composition, lifestyle, and environmental circumstances. Advances in genetics, combined with big data analytics, will enable individualized and targeted therapeutics. Telemedicine will broaden its scope, providing comprehensive and accessible healthcare solutions. Wearable gadgets and health-monitoring technology will allow people to take proactive control of their health.

3. Education and lifelong learning

The future of education places a premium on lifelong learning. Continuous technological innovation necessitates a workforce with adaptive capabilities. Online learning platforms, augmented reality (AR), and virtual reality (VR) will transform educational experiences by creating dynamic and interactive learning environments. Education will become more individualized, focusing on individual learning styles and interests.

4. Sustainable and green technologies

The broad use of green technology is critical for the future of sustainability. Renewable energy sources, such as solar and wind power, will dominate the energy landscape, helping to achieve a carbon-neutral future. Circular economy methods will reduce waste, and novel solutions will develop to address climate change issues. The Internet of Things (IoT) will play a critical role in developing smart, energy-efficient cities and infrastructure.

5. Space exploration and colonization

The future of space exploration promises to extend beyond the confines of our planet. Current missions to Mars and the Moon establish the framework for future human habitation. Private enterprises, in partnership with space agencies, will drive progress in space flight, making it more accessible.

The commercialization of space operations, such as asteroid mining and space tourism, will create new opportunities for exploration.

6. Blockchain and decentralization

Blockchain technology has a significant impact on the future of digital transactions and data management. Blockchain's decentralized and secure characteristics will change the way we handle financial transactions, maintain digital identities, and assure data integrity. Smart contracts will automate and streamline many procedures, hence minimizing the need for middlemen.

7. *Global Connectivity and 5G*

5G technology's broad adoption will change the future of communication. This high-speed, low-latency network will provide seamless communication, establishing the groundwork for the Internet of Things (IoT) to flourish. Smart cities, self-driving cars, and networked technologies will become ubiquitous in our daily lives, ushering in a hyper-connected society.

The future trajectories across industries provide a picture of a world full of opportunities. As we embrace technology developments, emphasize sustainability, and reimagine how we live, work, and connect, the future becomes a canvas for creativity and growth. Navigating the future demands a collaborative effort, a commitment to ethical concerns, and a proactive approach to resolving emerging difficulties. The decisions we make today will form the outlines of the future, and with each step into the unknown, we begin a journey toward a more interconnected, sustainable, and technologically sophisticated society.

CHAPTER 5

TECHNOLOGICAL INNOVATIONS AND FUNDING INITIATIVES

The dynamic panorama of technological innovation is fueled by a symbiotic link between ground-breaking ideas and the financial assistance necessary to make them a reality. In this inquiry, we will look at the junction of technical advancements and financing initiatives, emphasizing the critical role that financial backing plays in fostering the emergence of disruptive technologies across diverse industries.

1. The catalyst for innovation

Technological innovation is the heartbeat of development, igniting breakthroughs that reshape industries and improve human experiences. Researchers, scientists, and entrepreneurs generate innovative ideas in fields ranging from artificial intelligence to biotechnology. These visionaries find answers to challenging problems, bringing fresh ideas with the potential to change the way we live, work, and connect with the world.

2. Funding as the lifeblood

While breakthrough ideas lay the groundwork, money serves as the lifeblood that propels and nourishes these concepts into existence. In the lack of financial support, even the most revolutionary ideas may be limited to the realm of possibility. Funding provides the resources required for research, development, and the essential stages of transitioning technology from concept to commercialization.

3. Public and Private Investments

Wide ranges of financing sources influence the environment of technological innovation. Public investment, frequently assisted by government grants and programs, encourages research in areas having potential societal effects. Private investment, on the other hand, comes from venture capital companies, angel investors, and corporations looking to back new inventions and get a stake in their success.

4. Venture Capital and startups

One of the primary drivers of technical innovation is the symbiotic interaction between venture capital (VC) firms and startups. Venture capital businesses find high-potential entrepreneurs and provide financial resources to help them expand. In exchange, these businesses anticipate a return on investment, which contributes to the cycle of innovation and economic progress. Startups, renowned for their agility and revolutionary ideas, frequently rely on venture funding to overcome the difficult early phases of development.

5. Government Initiatives and Grants

Governments across the world understand the strategic necessity of encouraging innovation for economic growth and societal well-being. As a result, they provide major financial assistance for technical breakthroughs through grants, subsidies, and research initiatives. These programs frequently focus on critical sectors such as renewable energy, healthcare, and emerging technology to solve serious global concerns.

6. Corporate Research and Development

Large firms have an important role in encouraging innovation through significant investments in R&D. These expenditures help businesses to remain competitive, adapt to shifting market needs, and push new boundaries. Collaborations between companies and startups also create a mutually beneficial partnership in which startups get access to resources, mentoring, and market reach while corporations profit from new, nimble ideas.

7. Crowdfunding and Community Support

Crowdfunding systems, which allow anyone to donate modest sums to projects they believe in, have helped to democratize finance. This technique not only gives financial support but also assesses public interest and evaluates concepts before larger-scale expenditures. Community-driven support may be a powerful force, particularly for projects with a social effect or broad appeal.

In the complex dance of technological innovation and finance, collaboration among visionaries, investors, and supportive ecosystems generates a synergy that drives us into the future. As we see advances in domains such as artificial intelligence, biotechnology, and renewable energy, it becomes clear that long-term investment is critical for developing ideas into transformational technologies. The ever-changing landscape of technological innovation is being formed by the collaborative efforts of people who dare to dream, as well as the financial backing that makes these aspirations a reality. This interaction has the potential to address global concerns, spur economic progress, and create a future in which the frontiers of what is conceivable continue to expand.

CONCLUSION

In the complex worlds of technology, invention, and financing, a story of limitless possibility and collaborative growth emerges. From game-changing technology developments in healthcare and education to the transformational influence of sustainable practices, the interconnection of creative ideas and financial backing determines our common destiny.

As we investigate detailed insights into prospects across sectors, the recurring theme is flexibility, sustainability, and a dedication to pushing the frontiers of what is possible. Technological innovation, fuelled by a mutually beneficial connection with financing efforts, offers a future in which precision medicine, lifelong learning, renewable technology, and space exploration alter the possibilities available to us.

The coordinated efforts of academics, entrepreneurs, investors, and communities represent a dynamic environment in which growth is not only unavoidable, but also motivated by a common goal of a more interconnected, sustainable, and technologically sophisticated world. As we navigate the unknown, acknowledging the revolutionary power of technology advancements and the important role of financing, we begin on a journey in which each step forward adds to a future characterized by creativity, adaptation, and the never-ending search for a better tomorrow.

NOTES

NOTES

NOTES

NOTES

NOTES